AF879614

Ginseng Look-Alikes

Madison Woods

Wild Ozark, LLC

KINGSTON, AR

Copyright © 2016 by Madison Woods

All rights reserved. No part of this publication may be reproduced, distributed or transmitted in any form or by any means, without prior written permission.

Madison Woods/Wild Ozark, LLC
P.O. Box 173
Kingston, AR/72742
www.wildozark.com

The mission of Wild Ozark is to bridge the gap between people and the natural world, to help us remember that we are part of a larger whole.

Book Layout © 2017 BookDesignTemplates.com

Ginseng Look-Alikes/ Madison Woods. – 2nd ed.
ISBN 978-0-9961981-7-2

Dedicated to all the Nature Stewards out there

Contents

Ginseng Look-Alikes 1
Virginia creeper 3
Buckeye 9
Poison Ivy 13
Wild Strawberry 15
Elm Seedling 17
Hickory Sapling 19
Ginseng 21
ABOUT THE AUTHOR 23
Connect With Madison 23

CHAPTER ONE

Ginseng Look-Alikes

GINSENG IS NOTORIOUSLY HARD to spot in the wild. I still have a little trouble finding the first plant each season, even if I know there's one there because I planted it myself. But after spotting the first one, finding more is easier.

For those new to finding this plant there are a few plants that do a really good job of tricking the seeker into thinking they're finders.

Take heart, though. Once you become adept at telling the difference between ginseng (Panax Quinquefolius) and the other plants that look like it, you'll rarely make the mistake again. Eventually you'll wonder how you ever confused them with ginseng in the first place.

The following images and plant descriptions are of the most common look-alikes. There are other plants that some people find confusing, but these are the ones I am most often asked about.

Virginia creeper is the single most mistaken plant. Therefore the longest chapter will focus on that one.

Ideally you'll have examples of both plants in real life to look at. If you are able to travel to northwest Arkansas there are labeled plants at Compton Gardens in Bentonville and The Ozark Folk Center in Mountainview. Here at Wild Ozark in Kingston, I have a Ginseng Habitat Garden with plant markers to help people learn the differences plus I'm there to point them out in person.

Perhaps there is a botanical garden in your area or maybe you can find someone who knows the plants and will take you out to the woods to show you some examples.

Short of that, it is my hope that seeing the photos in this book will help you. As an added value, you are welcome to send me photos of your suspects and I'll let you know if what you've found is ginseng or a notorious look-alike. My email address is in the back.

This is a visual guide and less so an identification key. It is not intended to help you identify certain plants, per se, but to help you learn what is NOT ginseng and WHY it is not ginseng.

Lacy bug-eaten leaves of Virginia creeper in early fall.

CHAPTER TWO

Virginia creeper

Virginia creeper (Parthenocissus quinquefolia) has five leaflets at the end of a stem, just like ginseng. This look-alike is one of the most confusing and fools more people than any of the other plants.

The easiest way to identify on sight is that mature ginseng always has two ***significantly*** smaller leaves and three larger ones with the center of the larger ones being the longest. Sometimes creeper will exhibit two smaller leaves, but rarely to the same degree of difference as ginseng.

Another differences is the stem. Virginia creeper has a woody stem base. Ginseng does not. If you pull on the stem base of Virginia creeper it doesn't usually break off but will instead lift off the ground to reveal a vine. There are often small hairy rootlets attached to the vine in places where it makes contact with the soil. Ginseng will not do this. If you pull on a ginseng stem, it will break off, often at ground level.

The "prongs" of Virginia creeper are attached to a woody stem base often in more than one spot. Ginseng's prongs all come together and join at one point on a main stem that is connected to the root. On mature ginseng plants there will also be a flower or berry stalk attached at the junction where the prongs join.

Virginia creeper leaves are serrated. So are the leaves on ginseng. From plant to plant the serrations vary on the Virginia creeper. Younger plants often look different than older plants. Ginseng serrations are fairly constant from plant to plant, although the younger plants do indeed look different from the mature ones.

First look at the leaf shape. Ginseng always has a slightly elongated pointy tip with wider middles of each leaflet on a prong. Virginia creeper can also have elongated pointy tips, but very often they're more oval in shape.

Ginseng serrations (the "teeth", or indentations on the outer edges of a leaf) are longer toward the tip and often tiny toward the junction. Virginia creeper serrations vary, but are most of the time larger and more uniform throughout the leaflet except on the lower third. Most of the time, the creeper has no or very little serration from the junction where the leaf attaches to the stem to the upper 2/3 of the leaflet. Every once in a while, though, you'll find one that does have serrations all the way down.

When all of the visual clues are taken into consideration, you'll get better at determining the differences. After looking at the leaf shape, look at the stem. If you can't see the difference in the stem, pull the suspected creeper and see if it's attached by runners to other plants. If it is, then it's creeper. If it was ginseng and the stem breaks, the plant will regrow next year or go dormant a season and emerge the next. It won't kill the plant.

If you're patient and will watch the plant's behavior in late fall, Virginia creeper often turns bright red. Ginseng turns a sickly yellow. Virginia creeper plants use the same stem each year. Ginseng stems die back and whither with the first hard freeze and sometimes sooner.

Virginia creeper in early spring, younger leaves.

A mature creeper with smaller serrations, slightly different leaf shape.

Another Virginia creeper. You can see the woody stem base on this one.

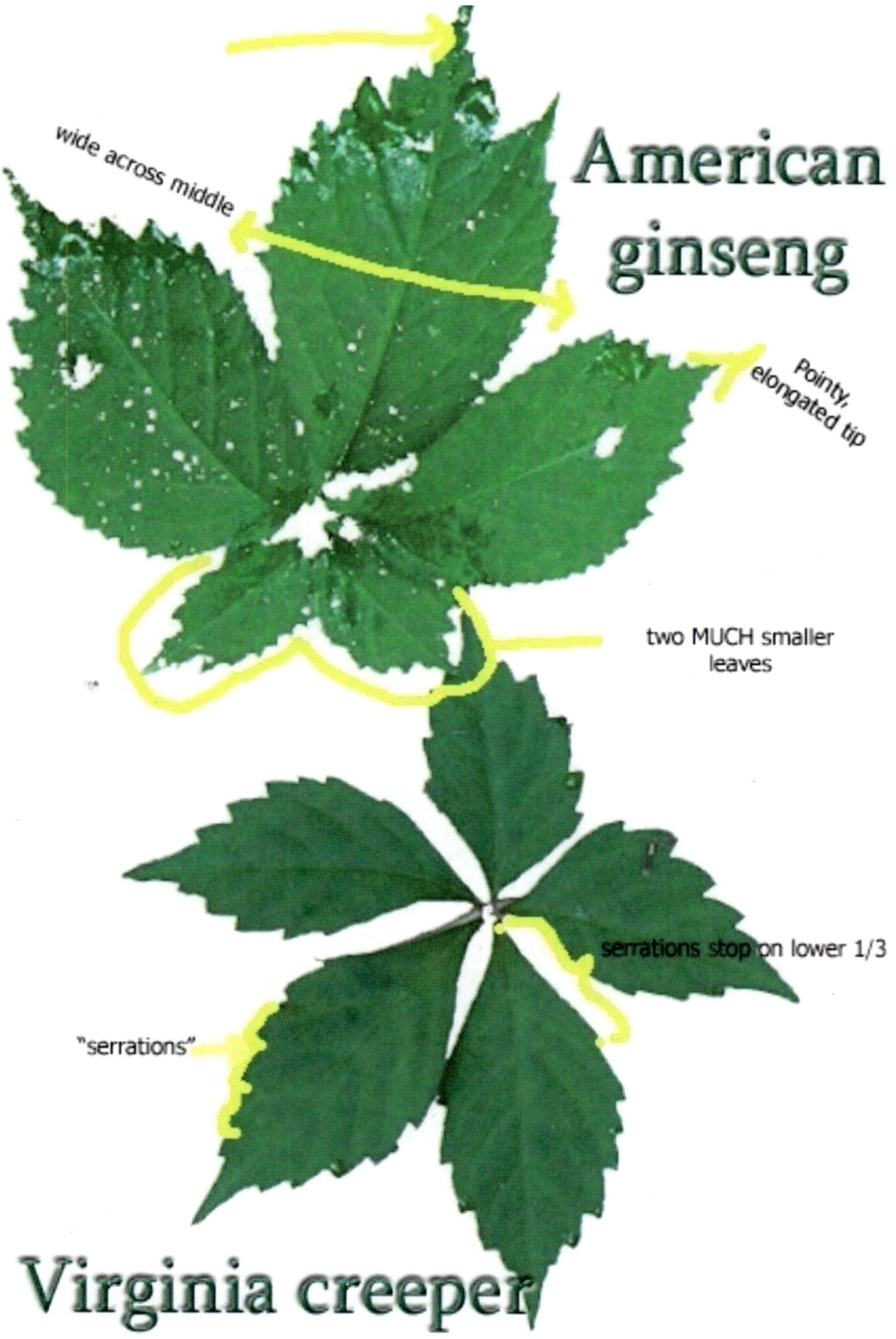
American ginseng
wide across middle
Pointy, elongated tip
two MUCH smaller leaves
serrations stop on lower 1/3
"serrations"
Virginia creeper

In fall buckeye turns red. In spring it produces red or yellow flower spires.

CHAPTER THREE

Buckeye

The buckeye is a tree with palmate leaves. The variety most commonly found here at Wild Ozark is the Ohio Buckeye (Aesculus glabra). In spring it makes pale yellow flowers arranged in spires on mature plants. Other buckeye varieties produce red flowers.

It's the seedlings that cause the confusion.

The leaves of buckeye have very fine serrations and prominent veins similar to ginseng. Sometimes you'll find buckeye seedlings that also have a couple of smaller leaves, but never to the degree of difference as ginseng's two smaller leaves.

Again, with this one, the stem is woody whereas ginseng's stem is not. If you pull on a buckeye seedling stem, you most likely will not pull up the plant but it takes a lot of tugging to break the stem free. You'll most likely strip the leaves off first.

If you tug on ginseng's stems, it will break off at the ground.

If you dig up a buckeye seedling, the root will be tough, fibrous, and woody. Ginseng roots have a consistency similar to carrots even if it does have a more complex shape than a carrot.

The buckeye uses the same stem year after year as it grows into a tree. Ginseng sends up new stems each year and the old ones die and whither each fall.

Ginseng is at the top, buckeye below.

Two more examples of buckeye seedlings.

Poison ivy
Ginseng
seedling

CHAPTER FOUR

Poison Ivy

This is not one I find similar enough to cause confusion, but many people do mistake it for ginseng. I get enough queries about it that I decided to include it.

It's mostly the seedlings of ginseng that bear some resemblance to poison ivy. Both have three leaves. Both are serrated.

But the ginseng serrations are smaller. Poison ivy is almost always connected to other plants by underground runners like the Virginia creeper.

If you pull up poison ivy, I'd suggest wearing gloves because the plant does cause allergic response in many people. In case you're not familiar with the reaction it causes, here's what happens. You'll get an itchy rash of small fluid-filled blisters. Some people are highly allergic and wind up in the emergency room for help to get it under control.

In late fall, poison ivy turns red just like the Virginia creeper and the buckeye. Ginseng seedlings just die back and whither into the leaf litter. Sometimes they'll turn pale or yellow first.

Wild strawberry and ginseng seedling side by side.

CHAPTER FIVE

Wild Strawberry

Sometimes wild strawberry leaves (Fragaria virginiana) are confused with ginseng seedlings.

The mature plants in fruit are easy enough to discern, but single leaves do have three leaves with serrations. A major difference is the shape of the leaf.

Wild strawberry leaves are rounded at the tip, whereas ginseng seedlings are pointed.

Ginseng seedlings have short stiff hairs on the surface of the leave along the central rib and strawberry does not.

Strawberry plants often are connected to surrounding strawberry plants by a trailing root similar to poison ivy and Virginia creeper. Ginseng is not.

The tiny little carrot shaped root of ginseng looks different than the spreading thin roots of strawberry.

If you dig up a strawberry or ginseng seedling to look at the roots, ginseng will have next year's stem bud already waiting for next season. It will be a small whitish root about an inch or half an inch long.

You can replant either one of these if you do decide to check below the surface.

Ginseng seedling at the top and bottom, elm seedling in the middle.

CHAPTER SIX

Elm Seedling

Elm seedlings look very much like ginseng seedlings. The easy way to differentiate is by noticing the amount of leaves. Elm seedlings have four and ginseng seedlings only have three.

If for some reason one of the leaves are missing from the elm, then you'd be able to tell from the roots. The elm will have a long and woody taproot, even if it is thin. The tiny ginseng seedling root is ready with next year's bud waiting at the base of the stem. This is not present on the elm root.

Hickory seedling/sapling.

CHAPTER SEVEN

Hickory Sapling

Nut tree saplings confuse a surprising amount of people. It's easy to see the difference once you notice where the two smaller leaves are attached to the stem.

Ginseng leaves are all attached at the same spot for each prong, and each prong is attached at the same spot on the stem.

After that, if there is still any doubt, take a look at the roots. You won't be able to just pull the hickory stem loose. It's usually firmly attached to the roots. If you dig it up, there will be woody, fibrous tree roots attached.

American ginseng first year seedling, second year seedling, and mature 4-prong with berries.

CHAPTER EIGHT

Ginseng

Now that we've taken a look at the most common look-alikes, it's time to take a look at American ginseng.

The facing page shows a seedling, a two year old plant, and a mature four-prong plant.

A "prong" is a set of leaves with a stem. What most people think of as the leaves are actually leaflets. The whole collection of leaves on a prong is the actual "leaf".

All five leaflets are attached at the same point on the prong. All prongs are attached at the same point on the stem.

At the base of the stem, where it attaches to the root, there is a leaf bud. This bud stays dormant until the following spring.

Next year's growth will start from that leaf bud. If you dig up roots to examine them during your observations take care to not break this bud.

If you do break it accidentally, the plant won't come up the following spring but it should not die. It will just stay dormant a year and emerge the following one.

Sometimes ginseng will stay dormant for no apparent reason and you won't see plants for a year or more and then suddenly they'll reappear as if nothing ever happened.

Good luck in your studies. Keep in mind that ginseng is an endangered plant with a fragile habitat. Be a good steward and only harvest when there are more than a few plants in a single area. Ideally you won't harvest when there are less than a hundred plants in a single colony, but colonies that large are a rare find these days.

If you make a conscious effort to leave at least half of the old mother plants behind and replant all of the seeds from the plants you harvest, there will be ginseng growing in your patch for generations to come.

I'm available to readers by email for questions. My address is madison@wildozark.com.

Happy 'Senging!

American ginseng
roots of various ages and sizes

ABOUT THE AUTHOR

Madison Woods is one half of the creative duo behind Wild Ozark. She writes and creates habitats for wild-simulated ginseng all over the woods on their property in northwest Arkansas.

Wild Ozark is perhaps the only certified American ginseng nursery in northwest Arkansas. They sell yearling seedlings from May through October to local customers from our Wild Ozark Nature Boutique in Kingston, AR and begin shipping bare-root in October.

You can read more about their retail outlet here: https://www.wildozark.com/nature-boutique/.

Visit www.wildozark.com for some homesteading, nature-loving blog posts.

Connect With Madison

Blog: http:/ /www.wildozark.com/blog/

Email: madison@wildozark.com

Newsletter: https:/ /www.wildozark.com/wild-ozark-musings/

Made in the USA
Monee, IL
05 November 2026

67948393R00019